YOUR KNOWLEDGE HAS VALUE

AF152140

- We will publish your bachelor's and master's thesis, essays and papers

- Your own eBook and book - sold worldwide in all relevant shops

- Earn money with each sale

Upload your text at www.GRIN.com and publish for free

Friederike Lange

Use of model organisms in Genetics

Student-directed learning (SDL)

GRIN Verlag

Bibliografische Information der Deutschen Nationalbibliothek:

Die Deutsche Bibliothek verzeichnet diese Publikation in der Deutschen National-
bibliografie; detaillierte bibliografische Daten sind im Internet über http://dnb.d-
nb.de/ abrufbar.

Imprint:

Copyright © 2010 GRIN Verlag GmbH
Druck und Bindung: Books on Demand GmbH, Norderstedt Germany
ISBN: 978-3-656-01614-4

This book at GRIN:

http://www.grin.com/en/e-book/179168/use-of-model-organisms-in-genetics

GRIN - Your knowledge has value

Der GRIN Verlag publiziert seit 1998 wissenschaftliche Arbeiten von Studenten, Hochschullehrern und anderen Akademikern als eBook und gedrucktes Buch. Die Verlagswebsite www.grin.com ist die ideale Plattform zur Veröffentlichung von Hausarbeiten, Abschlussarbeiten, wissenschaftlichen Aufsätzen, Dissertationen und Fachbüchern.

Visit us on the internet:

http://www.grin.com/

http://www.facebook.com/grincom

http://www.twitter.com/grin_com

Embedded in the course Genetics 3-H

SDL essay

Description and designing of experiments to address the unknown function of the novel gene *PUG1* in yeast (no obvious sequence homology with any known yeast gene; constantly expressed throughout cell cycle) following these questions:

Q1) How would you determine the function of *PUG1* in yeast?
Q2) How would you identify homologous *PUG1* gene(s) in one named invertebrate?
Q3) How would you determine the function of *PUG1* homologue(s) in this named invertebrate?
Q4) How would you identify homologous *PUG1* gene(s) in one named vertebrate?
Q5) How would you determine the function of *PUG1* homologue(s) in this named vertebrate?

Written by Friederike Lange

Group 6

16.12.2010

Q1) How would you determine the function of *PUG*1 in yeast?

"A gene is a discrete genomic region [...] which contains the information for the synthesis of functional proteins or non-coding RNAs"[1]. Thus the found gene sequence might be encoding for a protein or a non-coding RNA since it has an unknown function. "A non-coding RNA gene sequence does not have strong statistical signals, unlike protein coding genes"[2]. Thus, specific motifs/specific sequences give a clue about the gene product. The ORF (=open reading frame) ranges from the start codon (ATG) at one end to one of the three stop codons at the other end, with at least 100 bases in between. A gene encoding for a protein has normally specific additional sequences in and around the ORF, e.g. an enhancer, TATAboxes, a 5'UTR site, specific motifs like Leucine Zipper or Zinc Finger. For the following examination I assume that *PUG*1 encodes for a protein. With reverse genetics different approaches are feasible which can give clues about the function of *PUG*1.

So, to find the specific function of *PUG*1 there are two main approaches: One concerning the sequence itself and one concerning the function by introducing mutations.

The ORF from *PUG*1 was discovered through reassessing the known yeast genome sequence and it was found constantly expressed throughout all stages of the cell cycle. This implies that the protein is not involved in any specific phase of the cell cycle, or rather in its control system. But it does not mean that it is not needed in the cell cycle. It probably could be even needed for every step. Furthermore, no obvious homologous gene in yeast itself could be identified implying that the gene does not belong to a multi-family. Being homologous means that they are evolutionary linked by a common ancestor and thus, often (but not always) similar in sequence. Nevertheless, you should search for specific sequences (e.g. special motifs) giving a hint for the function. Additionally you could look on the genetic map for the surrounding genes which might indicate a linkage group functionally linked by specific sequences and consensus motifs. But the latter is probably not the case since no homologous sequences in yeast has been found. But in many cases, the structure has been highly conserved more than the amino acid sequence, since the function is more important. From the DNA sequence it is possible to predict the amino acid sequence and with that you could try to identify the 3D structure of the protein since some proteins are not similar in sequences but in structure. To accomplish this you could either do Circular Dichroism (CD) experiments/X-ray crystallography/Nuclear Magnetic Resonance or deduce the 3D structure on the basis of the amino acid sequence by structural superposition with special computer programmes. Comparing this 3D structure with other proteins from yeast could show a belonging to a family since related structures often imply related function.

In addition you could try to identify homologous gene sequences, protein sequences and 3D structures in other organisms by comparative genetics (covered by question 2-5), using sequence alignment programs like BLAST and FASTA; from their previously characterized functions you could also deduce the function of *PUG1*.

The found similarities could point in the correct direction, but it is still necessary to test these insights through direct experimentation.

To analyse/determine the function of *PUG1* it is possible to do a knockout experiment using homologous recombination (HR) so that they lack the gene or express an altered version of it. In HR you combine *PUG1* with a marker, like Kanamycin-resistance, and introduce this construct in the yeast genome, where it binds on the homologous strand and gets replicated. Then, observing the phenotype should give a hint of the *PUG1*'s function. I would look after obvious changes in shape and cell processes like growing and budding. In addition it is possible to introduce the yeast lacking *PUG1* to al lot of chemicals or to different stress factors. If the mutant differs, it shed light on *PUG1*'s function.

But if the gene is essential, a knockout mutant would be lethal precluding further studies of these mutants. An alternative would be to manipulate the gene in such a way, that it is over expressing and look if the phenotype differs or if it has an effect on other proteins.

Q2) How would you identify homologous *PUG1* gene(s) in one named invertebrate?

To identify homologous *PUG1* gene(s) in e.g. *Caenorhabditis elegans* it is possible to use a sequence alignment program like BLAST or FASTA to search for homologous genes in genomes of other organism or specifically of *C. elegans*. The program searches cluster of residues which fall into full or partial alignment. It is possible to search for an amino sequence and a nucleotide sequence which both could allow the prediction of *PUG1*'s function.

But even if there is a functionally similar protein, they could be too distantly related to be identified as clearly homologous in comparison to their amino acid sequences alone. Mutations will change the sequence of genes (largely without disruption their functions). Thus, a 3D structure prediction of *PUG1* from its amino acid sequence would improve the findings about its function from the sequence alignment. And since homology per definition (see page 1) does not exclusively mean that the sequences have to be similar, the structural comparisons would be a good additional way to find such dissimilar homologues. But since *PUG1* is unknown, a structure determination is relatively hard to work out. So, comparison of the amino acid sequences is the preferred way. To compare *PUG1* from yeast with that from *C. elegans* I would BLAST the sequence against the genome from *C. elegans* and score the number of matches. Thereby, you have to consider that the gene sequence likely has changed e.g. due to deletions, insertions, inversions and point mutations and the gene itself can be interrupted by diverse introns. Thus, you never get a complete homology and you have to allow gaps in the sequence alignments. The sequence alignment is then checked by generating randomly scrambled sequences (which also align and score) and look if the true sequence scores significantly better than the scrambled sequences. It is especially important by looking at the score alignments, that the 3'- , 5'-end regions and specific motifs giving the function are highly similar/highly conserved.

To omit the possibility that introns significantly interrupt the alignment, it would be better to compare cDNAs with each other. cDNA is made by copying the mRNA to DNA with reverse transcriptase, so that only the coding parts of the gene are in one row. I could either BLAST the cDNA of *PUG1* (which first must be produced) against the *C. elegans* genome or against the cDNA database of *C. elegans*. Also a library screening can be done through hybridisation test on a cDNA library of *C. elegans* to look if the *PUG1*-cDNA binds somewhere. You could also use degenerated cDNA probes of *PUG1* which do not bind too specifically, considering the changes in the sequences through evolution.

If you found a homologous gene in *C. elegans* by sequence alignment or with the aid of the other methods described above, you have to investigate its function through experiments, observing the mutant phenotypes and you could also try to rescue the mutant organisms with the *PUG1* gene of the other organism in both directions. If they get rescued the genes are definitely homologous (or rather orthologs, which are homologues in different organisms sharing the same ancestral gene). How this can be done, is described in the following question.

Q3) How would you determine the function of *PUG1* homologue(s) in this named invertebrate?

On the one hand the theoretical found sequence/protein/3D structure similarities described in Q2 give an evidence for the function of the *PUG1* homologue (=*PUG1c*) in *C. elegans*. On the other hand, provided that the methods in Q2 revealed a *PUG1c* in *C. elegans*, it is possible to determine the distinct function of *PUG1c* in *C. elegans* by perturbing the gene and producing mutants for *PUG1c*-loss-of-function as long as the gene mutant in yeast is not lethal. The observation of the phenotypes could give a clue for the function or could prove or disprove the theoretical function found in Q2.

Two main methods are imaginable: RNA interference (RNAi) and the creation of true deletion knockout mutants. RNAi is faster and therefore more appropriate for a rapid screen of changed phenotypes. It creates a knockdown of *PUG1c* by silencing its mRNA without altering the real DNA sequence. In contrast a creation of a deletion mutant needs more time, but would permanently remove all gene products. However, if the RNAi reveals a specific phenotype and a specific function is deductible, or even if no phenotype is observable a deletion mutant can be made to study the true knockout of *PUG1c*. But even if you do not find a specific phenotype, you should conduct a true knockout since RNAi does not work perfectly.

There are two ways for introducing an RNAi effect in *C. elegans*: Either injecting dsRNA of *PUG1c* into the germline of an adult hermaphrodite and observe the F1 offspring for a mutant phenotype (therefore, at first the dsRNA for *PUG1c* must be produced), or creating recombinant *E. coli* carrying a plasmid with *PUG1c* on which the worms must feed. For the latter you have to clone *PUG1c* in a strain of *E. coli*, where it gets expressed to dsRNA. The worms eat the bacteria and incorporate the dsRNA. An RNAi effect should be obtained both in the worms feeding on these *E. coli* and in their F1 offspring. The advantage of RNAi is especially that a lethal loss-of-function phenotype can be estimated. Furthermore, if *PUG1c* has redundant genes in *C. elegans*, RNAi would knock down all of them, so that the function of *PUG1c* cannot be replaced and the effects of the mutation could be studied.

If you have established the mutation and thus produced worm mutants in *PUG1c* you have to screen the offspring for changed phenotypes. If the phenotype is not obvious, you have to screen the worms under different conditions. If you got a clue of the function through the sequence alignments you can search specifically for this altered function.

However, to ensure the homologous relation between *PUG1* and *PUG1c*, it is possible to conduct a rescue experiments, either through complementation in a mutant *PUG1*-yeast with the *C. elegans'* wild type *PUG1c* gene, or via a transgenic rescue of a mutant PUG1c-*C. elegans* with the yeast wild type *PUG1* gene. For the latter you inject a solution of DNA vector molecules containing *PUG1* directly into the gonads of the adult hermaphrodites. The DNA molecules recombine with one another in the germ line nuclei and form large extra-chromosomal DNA molecules called "free arrays", which behave like mini-chromosomes. These free arrays are heritable and are transmitted through meiosis and mitosis. But only a few embryos will have incorporated the free arrays, which must be identified by an included appropriate marker in the transgenic DNA, e.g. GFP. If these worms do not show the previous mutant phenotype any longer and are phenotypical wild type again, the rescue was successful which implies that the two genes share the same function and are homologous.

Q4) How would you identify homologous *PUG1* gene(s) in one named vertebrate?

To identify homologous *PUG1* gene(s) in e.g. *Mus musculus*, again it is possible to use a sequence alignment program like BLAST or FASTA to search for homologous genes in the genome of *M. musculus*. As described in Q2 both programs search clusters of residues which fall into full or partial alignment. You should again search for an amino sequence and a nucleotide sequence similarity which both could allow the prediction of *PUG1*'s function. And now, since/if you have found a homologous gene in *C. elegans* as well, you can also try to BLAST the nucleotide and the peptide sequence of *PUG1c* against the mouse genome and look if you find homologues of such an organism which is evolutionary not as distant as yeast is to *M. musculus*. Because sequence divergence corresponds with evolutionary separation it is more likely to find a higher similarity score with *C. elegans* than with yeast.

But still, even if we find no clear homology between *PUG1/PUG1c* and mice genes, there could be a functionally similar protein, that does not any longer have a homologous sequence due to the evolutional changes. So if you just try to find homologues by comparing the DNA sequences with one another, it is very unlikely to find high similarities. Therefore, it is appropriate to compare not only the amino acid sequences but also the 3D structures. However, the comparison of the amino acid sequences is again the most convenient way. But it is to consider that the number and sizes of introns has increased in *M. musculus*. Therefore you should allow longer gaps in the alignments described in Q2.

You could also compare the cDNAs of *PUG1* and *PUG1c* with the mice cDNA library either per computer program or via a library screening per hybridisation test with *PUG1* on the *M. musculus* cDNA library. You should use degenerated probes from *PUG1* and *PUG1c*, which do not bind too specifically, considering the greater evolutionary gap and the higher amount of sequence alterations involved.

If you find a homologous gene in *M. musculus*, it can point your direction for the following necessary experiments to determine the function of this gene and looking for mutant phenotypes. Especially in mice it is more complicated to find a modified phenotype, if it is not a very obvious alteration. Due to the different tissues and well-defined compartments, finding changed gene expressions is not easy. An appropriate approach for an experiment for finding the function of a theoretical found homologue of *PUG1* in yeast or *PUG1c* in *C. elegans* is described in following Q5.

5) How would you determine the function of *PUG1* homologue(s) in this named vertebrate?

The possible function, revealed by the similarity comparisons (see Q4), are the basis now on which it is possible to try to identify the distinct function for the homologue of *PUG1* in *M. musculus* (= *PUG1m*) by conducting a knockout approach. In mice, genetic experiments are more complicated than in unicellular organisms. For producing animals that are transgenic in all cells we have to alter the embryo genes and transplant them into a foster mouse. Thereby, we have to keep in mind that in different tissues different genes can be expressed and that *PUG1m* is maybe just expressed in specific parts.

For producing a knockout mouse for *PUG1m* two transgenic approaches can be used: On the one hand it is possible to transform embryonic stem cells (ES cells) with DNA. On the other hand it is possible to inject DNA into the pronucleus of a fertilized mouse egg. In both methods you have to perform a targeted insertion, so that the wild type *PUG1m* gene gets deleted or replaced with another gene and thereby knocked out. Here it would be appropriate to replace *PUG1m* with e.g. GFP since we do not know the function of *PUG1m* and we would not know for what phenotypical change we should look to be sure, that the knockout had happened.

The targeted insertion involves homologous recombination, where two selectable markers are used to omit all non-recombinant or random recombinant cells: two appropriate markers are e.g. *neo* (an antibiotic resistance gene; positive marker) and *tk* (gene for an enzyme altering a nucleoside analogue which kills the cell by getting incorporated into replicating

DNA; negative marker). *Neo* shows that the recombinant DNA got inserted and *tk* ensures the targeted recombination (see Fig.1). So just the correctly inserted products survive and can be used for producing knockout offspring.

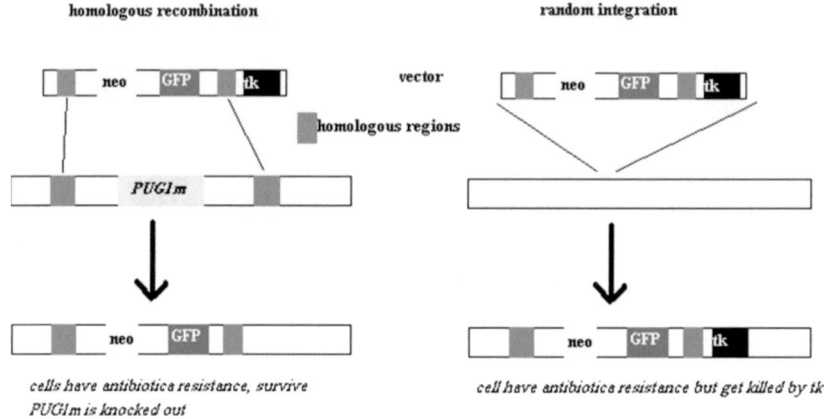

Figure 1: targeted insertion vs. random integration

For the first transgenic method you have to harvest ES cells from the inner cell mass (ICM) of a mouse blastocyst, which can grow in culture and still retain totipotent. You have to build the recombinant DNA, as described before, and transform ES cells with these vectors through e.g. electroporation. The ES cells, that incorporated the DNA (selected with the help of the markers), must be injected into the ICM of a mouse blastocyst which in turn has to be transferred in the uterus of a pseudo pregnant mouse. If they implant successfully it produces chimaera with wild type cells and recombinant/mutant cells. A recombinant sperm has to be taken and mated with a wild type in order to get true heterozygotes. You have to screen this offspring for the knockout by searching for GFP. Just some of them will be knocked out for *PUG*1m, and will be heterozygous for this knockout. By mating these individuals you get one homozygous *PUG*1m knockout individual of four and you can establish a homozygous transgene strain.

The second method for producing transgenic mice is similar to the first one. You have to harvest freshly fertilized eggs before the sperm head has become a pronucleus. The DNA (prepared like before) must be injected in the male pronucleus. After fusion to a diploid zygote nucleus and developing to a

2-cell embryo, these embryos get implanted in a pseudo pregnant foster mouse and the offspring has to be selected as before.

By investigating the knockout-GFP-mutant you can see where the *PUG1m* gene is normally expressed (through visualizing the expressed GFP) and you can observe the phenotypical changes occurring in these mutant mice. A screen for these changed phenotypes (if it's not obvious) has to be carried out by considering the single tissues.

Here as well, you could conduct a rescue experiment both in yeast and in mice which would show homologues function of *PUG1* and *PUG1m* if the rescue will be successful. For a rescue experiment in mice you could do the same transgenic approach as described above, but instead introducing GFP you introduce the wild type yeast *PUG1* gene and look if the mice phenotype get rescued.

References:

1) Pesole G; What is a gene? An updated operational definition, Gene 417 (2008), 1–4

2) Rivas E, Eddy SR; Noncoding RNA gene detecting using comparative sequence analysis, BMC Bioinformatics 2001, 2:8